RIVERS AND MOUNTAINS OF THE UNITED STATES

MW01526411

Contents

Chapter 1

Rivers and Mountains: An Introduction

The landscape of the United States includes hundreds of rivers and beautiful mountain ranges. People have the opportunity to hike and climb mountains. The features allow people to stay in shape while enjoying nature.

Did You Know?

There are 122 climbing gyms in the United States. People can practice their rock climbing skills at these gyms. Most climbing walls are 20 to 40 feet tall.

These **landforms**, which are found throughout the country, have played a role in American history. Both mountains and rivers have provided beauty and challenge to Americans throughout history. Consider the mountains and rivers in your region of the United States. In what ways have these landforms played a role in American history?

The Teton Mountain Range and Snake River in Wyoming.

Chapter 2

American Rivers

Rivers can flow into other rivers, lakes, or oceans. A river is a flowing stream of water. The water in rivers and streams is fresh water, not salty like the ocean.

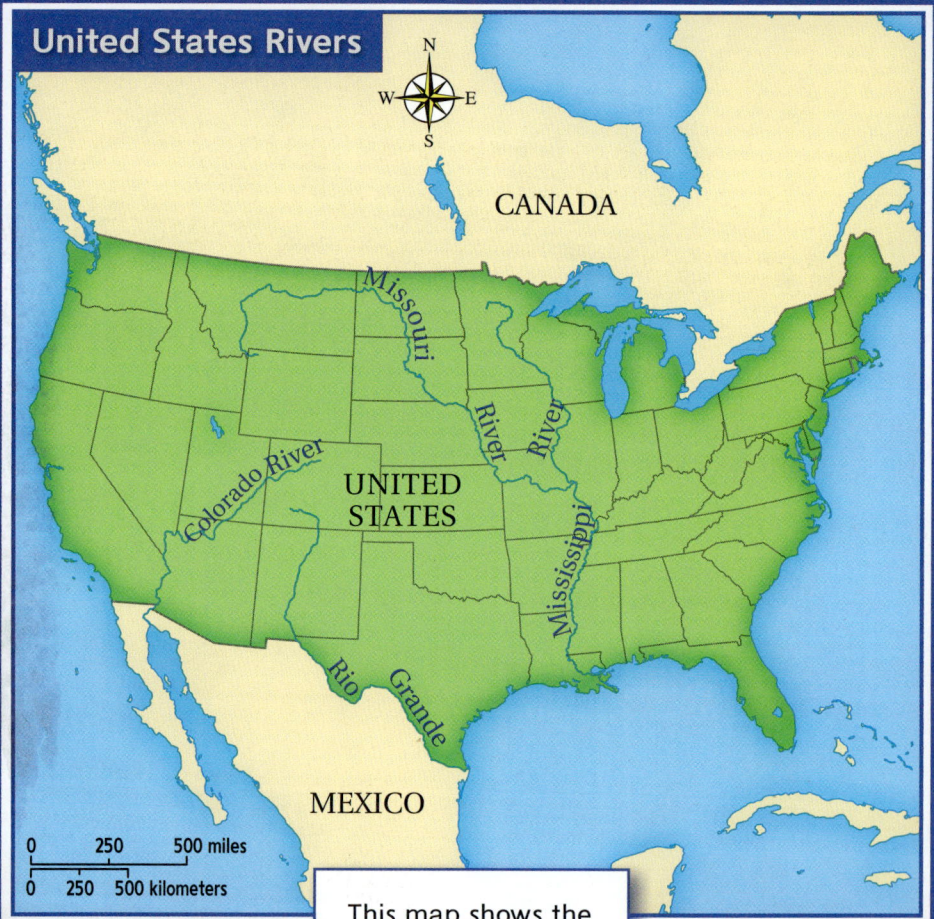

United States Rivers

This map shows the four major rivers in the United States.

Rivers have been vital to American history. In the 1700s and early 1800s, cars and trains had not yet been invented, so boats were the most efficient way to transport goods from one place to another. Many of the largest cities in the United States developed on rivers, along shipping centers. In addition, the land near the rivers provided good soil for growing food crops.

The Missouri River originates in Montana in the Rocky Mountains, flowing north and diverting southeast. The river is nicknamed "Big Muddy" because the soil it carries makes the water look brown. Just north of St. Louis, the Missouri River empties into the Mississippi River. The Missouri River's total length is 2,565 miles. Would 3,000 miles be a reasonable estimate of the length of Big Muddy?

Missouri River

At one time, Native Americans traded with French and Canadian explorers along the Missouri River. Years later, **settlers** followed the Missouri river on their journey to California. Many pioneers recognized the benefits of river transportation and settled towns close to it. However, when railroads began to ship goods through the region, the **population** around the river declined.

Did You Know?

The shortest river in the United States is Oregon's D River. It is only 120 feet long. The Missouri River is 2,565 miles long, which makes it the longest river. How can you use place value to show that the D River is much shorter than the Missouri River?

The Mississippi River flows 2,350 miles from Minnesota to Louisiana. Other rivers empty into the Mississippi. At the location where it meets the Missouri River and the Ohio River, it forms the third-largest river system in the world (3,877 miles). The distance of this entire river system is more than 1,000 miles farther than the distance from coast to coast.

Width of the Mississippi River At Different Points	
At Lake Itasca (MN)	30 feet
At Missouri River (MO)	3,500 feet
At Ohio River (IL)	4,500 feet
At Lake Onalaska (WI)	21,120 feet

Did You Know?

It would take 3 months for a raindrop falling into Lake Itasca to reach the Gulf of Mexico.

Early settlers used the Mississippi River as their primary trade **route.** However, the Civil War halted all trade; the river served as an important military travel route. Many battles took place on the river because both the North and the South wanted to control it.

Mississippi River in Iowa

TALK ABOUT IT

The table on page 8 shows the width of the Mississippi River at different points. What is the order of the widths from greatest to least?

Beginning in Colorado and ending at the Gulf of Mexico, the Rio Grande River is 1,885 miles long. Compare the Rio Grande River to the lengths of the Mississippi, Missouri, and D Rivers.

Native Americans used the Rio Grande River as a source of drinking water. They also used the river as an **irrigation** system as early as the 1500s.

Did You Know?

Rio Grande is Spanish for "big river."

The Rio Grande River functions as a border between Texas and Mexico. Many towns developed on either side of the river, in the United States and Mexico. As populations along it have increased, the river has become polluted. Water pollution is an issue for people who depend on the river to use it for drinking water and irrigation.

The Colorado River originates in the Rocky Mountains, and flows 1,450 miles to the Gulf of California. Crossing through three deserts on its route, this river has carved many canyons as it has flowed to the ocean. Can you name a famous canyon—it is also a national park—the Colorado River created?

Native Americans called the canyons of the Colorado River home for centuries. However, because it was difficult to navigate, Native Americans did not often travel on the Colorado River.

One Native American group, the Hohokam (huh hoh KAHM), built the largest early irrigation system in the West. Today 90% of Colorado River water is used for irrigation. In fact, the world's largest irrigation canal, the All-American Canal in California, carries water from the Colorado River. Every second, between 15,000 and 30,000 feet cubic of water passes through the canal. This volume of water could fill 11 average-sized swimming pools (20,000 gallons each) in 1 second!

Grand Canyon in Southwestern U.S.

The banks of the Colorado River are not highly populated. However, the dams on the Colorado River supply the cities of Los Angles, Las Vegas, San Diego, Phoenix, Tucson, and others with water. One of the world's largest dams, the Hoover Dam, provides power and flood control for the entire area. California and Arizona use water from the Colorado for irrigation.

TALK ABOUT IT

Is it fair to say that the Hoover Dam's height is greater than 750 feet? Is its width less than 1,250 feet?

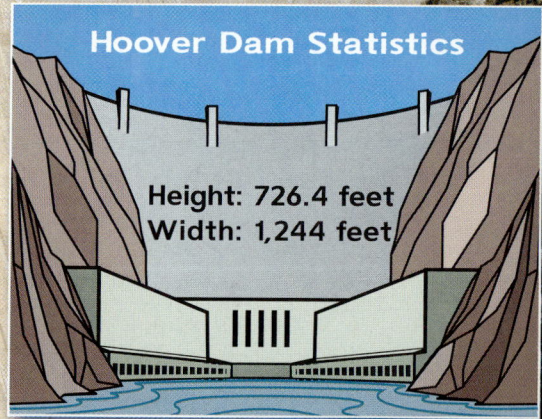

Hoover Dam Statistics

Height: 726.4 feet
Width: 1,244 feet

The Hoover Dam is located on the Nevada–Arizona border.

Chapter 3

American Mountains

America has many mountain ranges, with the majority in the West. The mountain ranges vary in both height and climate. These factors influenced how settlers **migrated** through the mountains. Elevation and climate continue to affect how people travel through the mountains.

United States Mountain Ranges

ROCKY MOUNTAINS

SIERRA NEVADA

APPALACHIAN MOUNTAINS

CANADA

UNITED STATES

MEXICO

PACIFIC OCEAN

ATLANTIC OCEAN

Gulf of Mexico

0 250 500 miles
0 250 500 kilometers

This map shows the three major mountain ranges in the United States.

The Rocky Mountain range extends about 3,000 miles from eastern Alaska to New Mexico. The highest peak is Mt. Elbert (14,431 feet above sea level), in Colorado. There are more than 100 smaller mountain ranges that comprise the Rocky Mountain range.

Climate in the Rocky Mountain Valleys	
Average annual temperature	43°F
Average temperature in July	82°F
Average temperature in January	7°F

Rocky Mountains

TALK ABOUT IT

How do these temperatures compare to those where you live?

Despite extreme elevations, there are passes through the Rocky Mountains. Settlers found that the South Pass in Wyoming, for example, made crossing the massive mountains possible. For those traveling through the south end of the range, the Santa Fe Trail led the way over the mountains.

Did You Know?

The Continental Divide is a ridge located along the top of the Rocky Mountains. Water on the east side of the Divide flows to the Atlantic Ocean. Water on the west side of the Divide flows to the Pacific Ocean.

As settlers traveled through the mountains, some decided to establish residence near natural resources: copper, silver, and gold, and abundant forests. As mining and forestry grew, the population of the Rocky Mountain region increased. Even today, the beautiful mountains attract visitors to the region.

Did You Know?

Rocky Mountain National Park is located in Colorado. The park was established in 1915. It has lakes, waterfalls, and more than 100 mountain peaks that are over 11,000 feet tall. The park covers 265,723 acres.

The Appalachian Mountains stretch from Canada to Alabama, about 1,600 miles. North Carolina has the highest peak, Mt. Mitchell (6,684 feet above sea level).

At one time, the Appalachians were a barrier for settlers heading west. Although the mountains in this range do not have the high elevations found in the western ranges, fewer passes allow for easy travel. The dense forests and ridges posed a risk for travelers. Native Americans, Spanish, and the French occupied the passes during the 1700s and 1800s. The difficult journey caused many settlers to remain in the mountains rather than to continue their travels. It did not take long before the settlers claimed most of the land.

TALK ABOUT IT

The Appalachian Trail is a hiking trail that runs from Maine to Georgia. It passes over the ridges of the Appalachian Mountains. The trail is 2,144 miles long. Would it take longer to hike the distance of this trail or the length of the Appalachian Mountain range?

The Appalachians are abundant with natural resources. Much of the area is rich in coal, iron, and gas. The dense forests' trees are valuable as lumber. But populations, terrain, and resources vary along this mountains range, producing an unstable economy in the Appalachians.

California's Sierra Nevada mountain range is a beautiful range, attracting tourists throughout the year. Located in eastern California, the range is about 400 miles long and between 40 and 80 miles wide. Mt. Whitney (14,495 feet above sea level) is the highest peak in the range.

Early settlers faced a difficult task. In spite of the high, jagged peaks and harsh winters, they persevered. The lure of gold was too tempting. Thousands of settlers managed to make their way through Sierras.

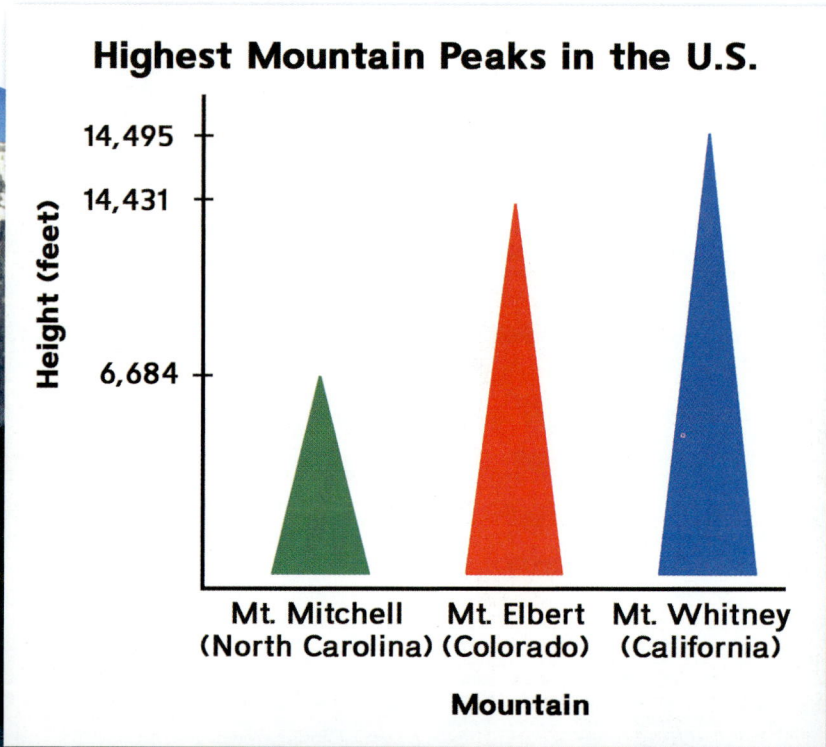

Highest Mountain Peaks in the U.S.

Height (feet)

Height	Mountain
14,495	
14,431	
6,684	

Mt. Mitchell (North Carolina) Mt. Elbert (Colorado) Mt. Whitney (California)

Mountain

Rivers and mountains decorate America's landscape. These landforms create a stunning environment in which to live, work, and play. They also affect the economy and determine populations.

Did You Know?

Mt. McKinley in Alaska is 20,320 feet high. How does its height compare to the other mountains that have been mentioned?

Glossary

irrigation

The use of ditches or pipes to bring water to dry land. *(page 10)*

landforms

Any of the shapes that make up Earth's surface. *(page 3)*

migrate

To move from one place to another. *(page 15)*

population

The number of people who live in a place or area. *(page 7)*

route

A path for traveling from one place to another. *(page 9)*

settlers

People that come to live in a new place. *(page 7)*